Copyright © 2011 XAMonline, Inc.
All rights reserved. No part of the material protected by this copyright notice may be reproduced or utilized in any form or by any means, electronic or mechanical, including photocopying, recording or by any information storage and retrievable system, without written permission from the copyright holder.

To obtain permission(s) to use the material from this work for any purpose including workshops or seminars, please submit a written request to:

XAMonline, Inc.
25 First Street, Suite 106
Cambridge, MA 02141
Toll Free: 1-800-509-4128
Email: info@xamonline.com
Web: www.xamonline.com
Fax: 1-617-583-5552

Library of Congress Cataloging-in-Publication Data

Wynne, Sharon A.
 Minnesota Middle Level Mathematics (5-8) Practice Test 1:
 Teacher Certification / Sharon A. Wynne. -1st ed.
 ISBN: 978-1-60787-285-6
  1. Minnesota Middle Level Mathematics (5-8) Practice Test 1
  2. Study Guides   3. MTLE   4. Teachers' Certification & Licensure
  5. Careers

Disclaimer:
The opinions expressed in this publication are the sole works of XAMonline and were created independently from the National Education Association, Educational Testing Service, or any State Department of Education, National Evaluation Systems or other testing affiliates.

Between the time of publication and printing, state specific standards as well as testing formats and website information may change that is not included in part or in whole within this product. Sample test questions are developed by XAMonline and reflect similar content as on real tests; however, they are not former tests. XAMonline assembles content that aligns with state standards but makes no claims nor guarantees teacher candidates a passing score. Numerical scores are determined by testing companies such as NES or ETS and then are compared with individual state standards. A passing score varies from state to state.

Printed in the United States of America   œ-1
Minnesota Middle Level Mathematics (5-8) Practice Test 1
ISBN: 978-1-60787-285-6

Middle School Mathematics
Pre-Test Sample Questions

DIRECTIONS: Read each item and select the best response.

1. $24 - 3 \times 7 + 2 =$
   (Average)

   A. 5

   B. 149

   C. –3

   D. 189

2. Which of the following sets is closed under division?
   (Average)

   A. Integers

   B. Rational numbers

   C. Natural numbers

   D. Whole numbers

3. Joe reads 20 words/min. and Jan reads 80 words/min. How many minutes will it take Joe to read the same number of words that it takes Jan 40 minutes to read?
   (Rigorous)

   A. 10

   B. 20

   C. 80

   D. 160

4. Choose the set in which the members are <u>not</u> equivalent.
   (Average)

   A. 1/2, 0.5, 50%

   B. 10/5, 2.0, 200%

   C. 3/8, 0.385, 38.5%

   D. 7/10, 0.7, 70%

5. Given W = whole numbers, N = natural numbers

   Z = integers
   R = rational numbers
   I = irrational numbers

   Which of the following is not true?
   (Easy)

   A. $R \subset I$

   B. $W \subset Z$

   C. $Z \subset R$

   D. $N \subset W$

6. Which denotes a complex number?
   (Rigorous)

   A. 3.678678678...

   B. $-\sqrt{27}$

   C. $123^{1/2}$

   D. $(-100)^{1/2}$

7. Express .0000456 in scientific notation.
   (Average)

   A. $4.56x10^{-4}$

   B. $45.6x10^{-6}$

   C. $4.56x10^{-6}$

   D. $4.56x10^{-5}$

8. Find the GCF of $2^2 \cdot 3^2 \cdot 5$ and $2^2 \cdot 3 \cdot 7$.
   (Average)

   A. $2^5 \cdot 3^3 \cdot 5 \cdot 7$

   B. $2 \cdot 3 \cdot 5 \cdot 7$

   C. $2^2 \cdot 3$

   D. $2^3 \cdot 3^2 \cdot 5 \cdot 7$

9. Which of the following is always composite if x is odd, y is even, and both x and y are greater than or equal to 2?
   (Rigorous)

   A. $x+y$

   B. $3x+2y$

   C. $5xy$

   D. $5x+3y$

10. If three cups of concentrate are needed to make 2 gallons of fruit punch, how many cups are needed to make 5 gallons?
    (Easy)

    A. 6 cups

    B. 7 cups

    C. 7.5 cups

    D. 10 cups

11. The volume of water flowing through a pipe varies directly with the square of the radius of the pipe. If the water flows at a rate of 80 liters per minute through a pipe with a radius of 4 cm, at what rate would water flow through a pipe with a radius of 3 cm?
    (Average)

    A. 45 liters per minute

    B. 6.67 liters per minute

    C. 60 liters per minute

    D. 4.5 liters per minute

12. Simplify $\dfrac{\frac{3}{4}x^2y^{-3}}{\frac{2}{3}xy}$

    (Rigorous)

    A. $\frac{1}{2}xy^{-4}$

    B. $\frac{1}{2}x^{-1}y^{-4}$

    C. $\frac{9}{8}xy^{-4}$

    D. $\frac{9}{8}xy^{-2}$

13. Solve: $\sqrt{75} + \sqrt{147} - \sqrt{48}$
    (Rigorous)

    A. 174

    B. $12\sqrt{3}$

    C. $8\sqrt{3}$

    D. 74

14. Solve for x:
    $3x + 5 \geq 8 + 7x$
    (Average)

    A. $x \geq -\frac{3}{4}$

    B. $x \leq -\frac{3}{4}$

    C. $x \geq \frac{3}{4}$

    D. $x \leq \frac{3}{4}$

15. What is the solution set for the following equations?
    (Rigorous)

    $3x + 2y = 12$
    $12x + 8y = 15$

    A. All real numbers

    B. $x = 4, y = 4$

    C. $x = 2, y = -1$

    D. $\varnothing$

16. **Identify the proper sequencing of subskills when teaching graphing inequalities in two dimensions**
    (Easy)

    A. Shading regions, graphing lines, graphing points, determining whether a line is solid or broken

    B. Graphing points, graphing lines, determining whether a line is solid or broken, shading regions

    C. Graphing points, shading regions, determining whether a line is solid or broken, graphing lines

    D. Graphing lines, determining whether a line is solid or broken, graphing points, shading regions

17. **Factor completely:**
    8(x – y) + a(y – x)
    (Average)

    A. (8 + a) (y – x)

    B. (8 – a) (y – x)

    C. (a – 8) (y – x)

    D. (a – 8) (y + x)

18. **What would be the seventh term of the expanded binomial $(2a+b)^8$?**
    (Rigorous)

    A. $2ab^7$

    B. $41a^4b^4$

    C. $112a^2b^6$

    D. $16ab^7$

19. **Solve for x:**
    |2x +3| > 4
    (Rigorous)

    A. $-\frac{7}{2} > x > \frac{1}{2}$

    B. $-\frac{1}{2} > x > \frac{7}{2}$

    C. x < $\frac{7}{2}$ or x<$-\frac{1}{2}$

    D. x<$-\frac{7}{2}$ or x>$\frac{1}{2}$

20. **Given a 30-by-60-meter garden with a circular fountain with a 5 meter radius, calculate the area of the portion of the garden not occupied by the fountain.**
    (Rigorous)

    A. 1721 m²

    B. 1879 m²

    C. 2585 m²

    D. 1015 m²

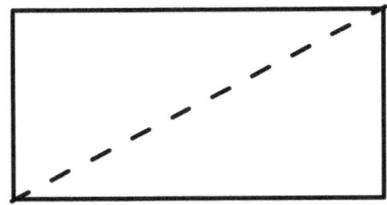

21. The above diagram is most likely used in deriving a formula for which of the following?
    (Easy)

    A. The area of a rectangle

    B. The area of a triangle

    C. The perimeter of a triangle

    D. The surface area of a prism

22. Determine the volume of a sphere to the nearest cm if the surface area is 113 cm². *(Rigorous)*

    A. 113 cm³

    B. 339 cm³

    C. 37.7 cm³

    D. 226 cm3

23. Which is a postulate?
    (Easy)

    A. The sum of the angles in any triangle is 180°

    B. A line intersects a plane in one point

    C. Two intersecting lines form congruent vertical angles

    D. Any segment is congruent to itself

24. Given similar polygons with corresponding sides of lengths 9 and 15, find the perimeter of the smaller polygon if the perimeter of the larger polygon is 150 units.
    (Rigorous)

    A. 54

    B. 135

    C. 90

    D. 126

25. Given $l_1 \parallel l_2$ prove $\angle b \cong \angle e$

   1) $\angle b \cong \angle d$ 1) vertical angle theorem

   2) $\angle d \cong \angle e$ 2) alternate interior angle theorem

   3) $\angle b \cong \angle e$ 3) symmetric axiom of equality

   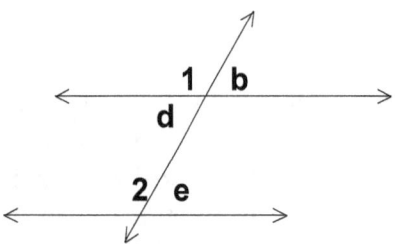

   Which step is incorrectly justified?
   (Average)

   A. Step 1

   B. Step 2

   C. Step 3

   D. No error

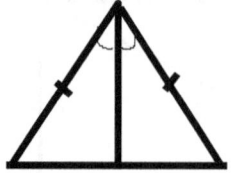

26. What method could be used to prove the above triangles congruent?
    (Average)

    A. SSS

    B. SAS

    C. AAS

    D. SSA

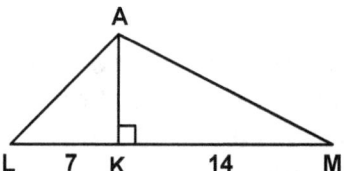

27. Given altitude AK in the right triangle ALM with measurements as indicated, determine the length of AK.
    (Rigorous)

    A. 98

    B. $7\sqrt{2}$

    C. $\sqrt{21}$

    D. $7\sqrt{3}$

28. Line *p* has a negative slope and passes through the point (0, 0). If line *q* is perpendicular to line *p*, which of the following must be true?
    (Rigorous)

    A. Line *q* has a negative *y*-intercept

    B. Line *q* passes through the point (0,0)

    C. Line *q* has a positive slope

    D. Line *q* has a positive *y*-intercept

29. Find the distance between (3,7) and (−3,4).
    (Average)

    A. 9

    B. 45

    C. $3\sqrt{5}$

    D. $5\sqrt{3}$

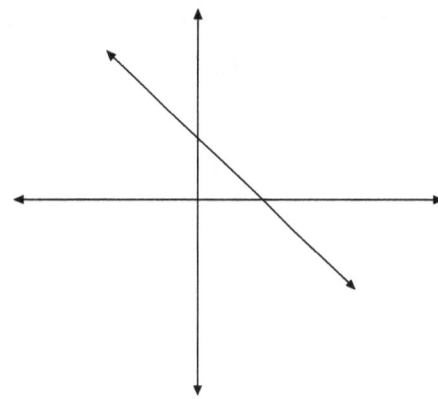

30. Which equation is represented by the above graph?
    (Average)

    A. x − y = 3

    B. x − y = −3

    C. x + y = 3

    D. x + y = −3

31. Give the domain for the function over the set of real numbers:

    $$y = \frac{3x+2}{2x^2-3}$$

    (Rigorous)

    A. All real numbers

    B. All real numbers, x ≠ 0

    C. All real numbers, x ≠ −2 or 3

    D. All real numbers, x ≠ $\frac{\pm\sqrt{6}}{2}$

32. Which statement is true about George's budget?
    (Easy)

    A. George spends the greatest portion of his income on food

    B. George spends twice as much on utilities as he does on his mortgage

    C. George spends twice as much on utilities as he does on food

    D. George spends the same amount on food and utilities as he does on mortgage

33. Given a spinner with the numbers one through eight, what is the probability that you will spin an even number or a number greater than four?
    (Easy)

    A. 1/4

    B. 1/2

    C. 3/4

    D. 1

34. Given a drawer with 5 black socks, 3 blue socks, and 2 red socks, what is the probability that you will draw two black socks in two draws in a dark room?
    (Average)

    A. 2/9

    B. 1/4

    C. 17/18

    D. 1/18

35. If a horse will probably win three races out of ten, what are the odds that he will win?
    (Rigorous)

    A. 3:10

    B. 7:10

    C. 3:7

    D. 7:3

36. Find the median of the following set of data:

    14   3   7   6   11   20

    (Average)

    A. 9

    B. 8.5

    C. 7

    D. 11

37. Corporate salaries are listed for several employees. Which would be the best measure of central tendency?
    (*Average*)

    $24,000 $24,000 $26,000
    $28,000 $30,000 $120,000

    A. Mean

    B. Median

    C. Mode

    D. No difference

38. A student scored in the 87th percentile on a standardized test. Which would be the best interpretation of his score?
    (*Easy*)

    A. Only 13% of the students who took the test scored higher

    B. This student should be getting mostly Bs on his report card

    C. This student performed below average on the test

    D. This is the equivalent of missing 13 questions on a 100 question exam

39. Determine the number of subsets of set K.

    K = {4, 5, 6, 7}

    (*Average*)

    A. 15

    B. 16

    C. 17

    D. 18

40. {6, 11, 16, 21 . . .}

    Find the sum of the first 20 terms in the sequence.
    (*Rigorous*)

    A. 1070

    B. 1176

    C. 969

    D. 1069

Middle School Mathematics
Pre-Test Sample Questions with Rationales

The following represent one way to solve the problems and obtain a correct answer. There are many other mathematically correct ways of determining the correct answer.

1. $24 - 3 \times 7 + 2 =$
   (*Average*)

   A. 5

   B. 149

   C. −3

   D. 189

Answer: A. 5
According to the order of operations, multiplication is performed first, and then addition and subtraction from left to right.

2. Which of the following sets is closed under division?
   (*Average*)

   A. Integers

   B. Rational numbers

   C. Natural numbers

   D. Whole numbers

Answer: B. Rational numbers
In order to be closed under division, when any two members of the set are divided, the answer must be contained in the set. This is not true for integers, natural, or whole numbers as illustrated by the counter example 11/2 = 5.5.

3. Joe reads 20 words/min. and Jan reads 80 words/min. How many minutes will it take Joe to read the same number of words that it takes Jan 40 minutes to read?
   (Rigorous)

   A.   10

   B.   20

   C.   80

   D.   160

Answer: D. 160
If Jan reads 80 words/minute, she will read 3200 words in 40 minutes.
$\frac{3200}{20} = 160$ At 20 words per minute, it will take Joe 160 minutes to read 3200 words.

4. Choose the set in which the members are <u>not</u> equivalent.
   (Average)

   A.   1/2, 0.5, 50%

   B.   10/5, 2.0, 200%

   C.   3/8, 0.385, 38.5%

   D.   7/10, 0.7, 70%

Answer: C. 3/8, 0.385, 38.5%
3/8 is equivalent to .375 and 37.5%.

5. Given W = whole numbers, N = natural numbers,

   Z = integers
   R = rational numbers
   I = irrational numbers

   Which of the following is not true?
   (Easy)

   A.   R ⊂ I

   B.   W ⊂ Z

   C.   Z ⊂ R

   D.   N ⊂ W

Answer: A. R ⊂ I
The rational numbers are not a subset of the irrational numbers. All of the other statements are true.

6. **Which denotes a complex number?**
   (Rigorous)

   A.   3.678678678…

   B. $-\sqrt{27}$

   C. $123^{1/2}$

   D. $(-100)^{1/2}$

Answer: D. $(-100)^{1/2}$
A complex number is the square root of a negative number. The complex number is defined as the square root of –1. The exponent ½ represents a square root.

7. **Express .0000456 in scientific notation.**
   (Average)

   A. $4.56 x 10^{-4}$

   B. $45.6 x 10^{-6}$

   C. $4.56 x 10^{-6}$

   D. $4.56 x 10^{-5}$

Answer: D. $4.56 x 10^{-5}$
In scientific notation, the decimal point belongs to the right of the 4, the first significant digit. To get from 4.56×10^{-5} back to 0.0000456, we would move the decimal point 5 places to the left.

8. **Find the GCF of $2^2 \cdot 3^2 \cdot 5$ and $2^2 \cdot 3 \cdot 7$.**
   (Average)

   A. $2^5 \cdot 3^3 \cdot 5 \cdot 7$

   B. $2 \cdot 3 \cdot 5 \cdot 7$

   C. $2^2 \cdot 3$

   D. $2^3 \cdot 3^2 \cdot 5 \cdot 7$

Answer: C. $2^2 \cdot 3$
Choose the number of each prime factor that is in common.

9. Which of the following is always composite if x is odd, y is even, and both x and y are greater than or equal to 2?
   (Rigorous)

   A. $x+y$

   B. $3x+2y$

   C. $5xy$

   D. $5x+3y$

Answer: C. $5xy$

A composite number is a number, which is not prime. The prime number sequence begins 2, 3, 5, 7, 11, 13, 17... To determine which of the expressions is <u>always</u> composite, experiment with different values of x and y, such as x=3 and y=2, or x=5 and y=2. It turns out that 5xy will always be an even number, and therefore, composite, if y=2.

10. If three cups of concentrate are needed to make 2 gallons of fruit punch, how many cups are needed to make 5 gallons?
    (Easy)

    A.   6 cups

    B.   7 cups

    C.   7.5 cups

    D.   10 cups

Answer: C. 7.5 cups
Set up the proportion 3/2 = x/5, cross multiply to obtain 15=2x, then divide both sides by 2.

11. The volume of water flowing through a pipe varies directly with the square of the radius of the pipe. If the water flows at a rate of 80 liters per minute through a pipe with a radius of 4 cm, at what rate would water flow through a pipe with a radius of 3 cm?
    (Average)

    A. 45 liters per minute

    B. 6.67 liters per minute

    C. 60 liters per minute

    D. 4.5 liters per minute

Answer: A. 45 liters per minute
Set up the direct variation: $\frac{V}{r^2} = \frac{V}{r^2}$. Substituting gives $\frac{80}{16} = \frac{V}{9}$. Solving for V gives 45 liters per minute.

12. Simplify $\dfrac{\frac{3}{4} x^2 y^{-3}}{\frac{2}{3} xy}$

    (Rigorous)

    A. $\frac{1}{2} xy^{-4}$

    B. $\frac{1}{2} x^{-1} y^{-4}$

    C. $\frac{9}{8} xy^{-4}$

    D. $\frac{9}{8} xy^{-2}$

Answer: C. $\frac{9}{8} xy^{-4}$
Simplify the complex fraction by inverting the denominator and multiplying: 3/4(3/2) =9/8, then subtract exponents to obtain the correct answer.

13. **Solve:** $\sqrt{75} + \sqrt{147} - \sqrt{48}$
    (Rigorous)

    A. 174

    B. $12\sqrt{3}$

    C. $8\sqrt{3}$

    D. 74

Answer: C. $8\sqrt{3}$
Simplify each radical by factoring out the perfect squares: $5\sqrt{3} + 7\sqrt{3} - 4\sqrt{3} = 8\sqrt{3}$

14. **Solve for x:**
    $3x + 5 \geq 8 + 7x$
    (Average)

    A. $x \geq -\frac{3}{4}$

    B. $x \leq -\frac{3}{4}$

    C. $x \geq \frac{3}{4}$

    D. $x \leq \frac{3}{4}$

Answer: B. $x \leq -\frac{3}{4}$
Using additive equality, $-3 \geq 4x$. Divide both sides by 4 to obtain $-3/4 \geq x$. Carefully determine which answer choice is equivalent.

15. **What is the solution set for the following equations?**
    (Rigorous)

    $3x + 2y = 12$
    $12x + 8y = 15$

    A. All real numbers

    B. $x = 4, y = 4$

    C. $x = 2, y = -1$

    D. $\emptyset$

Answer: D. $\emptyset$
Multiplying the top equation by −4 and adding results in the equation 0 = −33. Since this is a false statement, the correct choice is the null set.

16. **Identify the proper sequencing of subskills when teaching graphing inequalities in two dimensions**
    (Easy)

    A. Shading regions, graphing lines, graphing points, determining whether a line is solid or broken

    B. Graphing points, graphing lines, determining whether a line is solid or broken, shading regions

    C. Graphing points, shading regions, determining whether a line is solid or broken, graphing lines

    D. Graphing lines, determining whether a line is solid or broken, graphing points, shading regions

Answer: B. Graphing points, graphing lines, determining whether a line is solid or broken, shading regions

17. Factor completely:

    8(x – y) + a(y – x)

    (*Average*)

    A.   (8 + a) (y – x)

    B.   (8 – a) (y – x)

    C.   (a – 8) (y – x)

    D.   (a – 8) (y + x)

Answer: C. (a – 8) (y – x)
Glancing first at the solution choices, factor (y – x) from each term. This leaves –8 from the first reran and a from the send term: (a – 8) (y – x)

18. What would be the seventh term of the expanded binomial $(2a+b)^8$?
    (*Rigorous*)

    A. $2ab^7$

    B. $41a^4b^4$

    C. $112a^2b^6$

    D. $16ab^7$

Answer: C. $112a^2b^6$

The set-up for finding the seventh term is $\frac{8(7)(6)(5)(4)(3)}{6(5)(4)(3)(2)(1)}(2a)^{8-6}b^6$, which gives $28(4a^2b^6)$ or $112a^2b^6$. One can also find the right answer simply by looking at the powers of a and b.

19. **Solve for x:**
    $|2x+3| > 4$
    (Rigorous)

    A. $\quad -\frac{7}{2} > x > \frac{1}{2}$

    B. $\quad -\frac{1}{2} > x > \frac{7}{2}$

    C. $\quad x < \frac{7}{2}$ or $x < -\frac{1}{2}$

    D. $\quad x < -\frac{7}{2}$ or $x > \frac{1}{2}$

Answer: D. $x < -\frac{7}{2}$ or $x > \frac{1}{2}$
The quantity within the absolute value symbols must be either > 4 or < –4. Solve the two inequalities 2x + 3 > 4 or 2x + 3 < –4

20. **Given a 30-by-60-meter garden with a circular fountain with a 5 meter radius, calculate the area of the portion of the garden not occupied by the fountain.**
    (Rigorous)

    A. $\quad$ 1721 m²

    B. $\quad$ 1879 m²

    C. $\quad$ 2585 m²

    D. $\quad$ 1015 m²

Answer: A. 1721 m²
Find the area of the garden and then subtract the area of the fountain: 30(60)– $\pi(5)^2$ or approximately 1721 square meters.

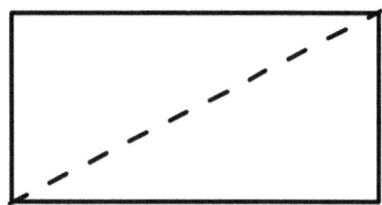

21. The above diagram is most likely used in deriving a formula for which of the following?
    (Easy)

    A. The area of a rectangle

    B. The area of a triangle

    C. The perimeter of a triangle

    D. The surface area of a prism

Answer: B. The area of a triangle

22. Determine the volume of a sphere to the nearest cm if the surface area is 113 cm².
    (*Rigorous*)

    A. 113 cm³

    B. 339 cm³

    C. 37.7 cm³

    D. 226 cm³

Answer: A. 113 cm³
Solve for the radius of the sphere using $A = 4\Pi r^2$. The radius is 3. Then, find the volume using $4/3\ \Pi r^3$. Only when the radius is 3 are the volume and surface area equivalent.

23. **Which is a postulate?**
    (Easy)

    A. The sum of the angles in any triangle is 180°.

    B. A line intersects a plane in one point.

    C. Two intersecting lines form congruent vertical angles.

    D. Any segment is congruent to itself.

Answer: D. Any segment is congruent to itself.
A postulate is an accepted property of real numbers or geometric figures, which cannot be proven; choices A, B. and C are theorems which can be proven.

24. **Given similar polygons with corresponding sides of lengths 9 and 15, find the perimeter of the smaller polygon if the perimeter of the larger polygon is 150 units.**
    (Rigorous)

    A. 54

    B. 135

    C. 90

    D. 126

Answer: C. 90
The perimeters of similar polygons are directly proportional to the lengths of their sides, therefore 9/15 = x/150. Cross-multiply to obtain 1350 = 15x, then divide by 15 to obtain the perimeter of the smaller polygon.

25. Given $l_1 \parallel l_2$ prove $\angle b \cong \angle e$

1) $\angle b \cong \angle d$ 1) vertical angle theorem
2) $\angle d \cong \angle e$ 2) alternate interior angle theorem
3) $\angle b \cong \angle 3$ 3) symmetric axiom of equality

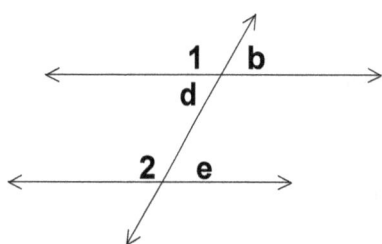

Which step is incorrectly justified?
(Average)

A. Step 1

B. Step 2

C. Step 3

D. No error

Answer: C. Step 3
Step 3 can be justified by the transitive property.

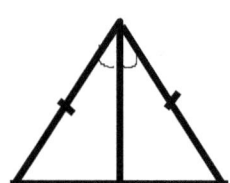

26. **What method could be used to prove the above triangles congruent?**
    (Average)

    A.   SSS

    B.   SAS

    C.   AAS

    D.   SSA

Answer: B. SAS
Use SAS with the last side being the vertical line common to both triangles.

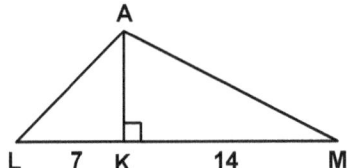

27. **Given altitude AK in the right triangle ALM with measurements as indicated, determine the length of AK.**
    (Rigorous)

    A.   98

    B. $7\sqrt{2}$

    C. $\sqrt{21}$

    D. $7\sqrt{3}$

Answer: B. $7\sqrt{2}$
The attitude from the right angle to the hypotenuse of any right triangle is the geometric mean of the two segments that are formed. Multiply 7 x 14 and take the square root.

28. Line *p* has a negative slope and passes through the point (0, 0). If line *q* is perpendicular to line *p*, which of the following must be true? *(Rigorous)*

    A.   Line *q* has a negative *y*-intercept

    B.   Line *q* passes through the point (0,0)

    C.   Line *q* has a positive slope

    D.   Line *q* has a positive *y*-intercept

Answer: C. Line *q* has a positive slope.
Draw a picture to help you visualize the problem.

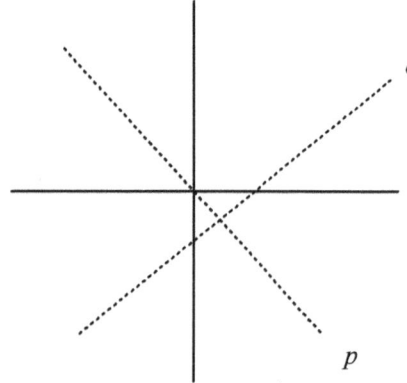

Choices A and D are not correct because line *q* could have a positive or a negative *y*-intercept. Choice B is incorrect because line *q* does not necessarily pass through (0, 0). Since line *q* is perpendicular to line *p*, which has a negative slope, it must have a positive slope.

29. **Find the distance between (3,7) and (−3,4).**
    (Average)

    A. 9

    B. 45

    C. $3\sqrt{5}$

    D. $5\sqrt{3}$

Answer: C. $3\sqrt{5}$
Using the distance formula
$$\sqrt{[3-(-3)]^2 + (7-4)^2}$$
$$= \sqrt{36+9} = \sqrt{45} = 3\sqrt{5}$$

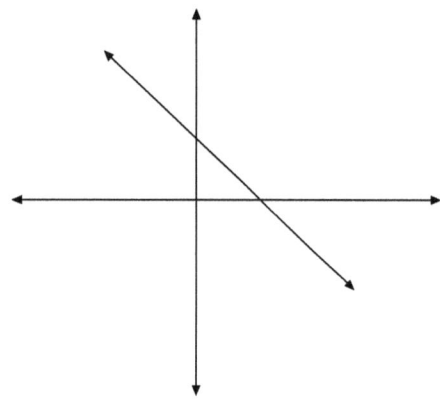

30. Which equation is represented by the above graph?
    (*Average*)

    A. $x - y = 3$

    B. $x - y = -3$

    C. $x + y = 3$

    D. $x + y = -3$

Answer: C. $x + y = 3$

By looking at the graph, we can determine the slope to be −1 and the y-intercept to be 3. Write the slope intercept form of the line as y = −1x + 3. Add x to both sides to obtain x + y = 3, the equation in standard form.

31. Give the domain for the function over the set of real numbers:

$$y = \frac{3x+2}{2x^2-3}$$

(Rigorous)

- A. all real numbers
- B. all real numbers, $x \neq 0$
- C. all real numbers, $x \neq -2$ or 3
- D. all real numbers, $x \neq \frac{\pm\sqrt{6}}{2}$

Answer: D. all real numbers, $x \neq \frac{\pm\sqrt{6}}{2}$

Solve the denominator for 0. These values will be excluded from the domain.

$$2x^2 - 3 = 0$$
$$2x^2 = 3$$
$$x^2 = 3/2$$

$$x = \sqrt{\tfrac{3}{2}} = \sqrt{\tfrac{3}{2}} \cdot \sqrt{\tfrac{2}{2}} = \frac{\pm\sqrt{6}}{2}$$

32. Which statement is true about George's budget?
    (Easy)

    A. George spends the greatest portion of his income on food.

    B. George spends twice as much on utilities as he does on his mortgage.

    C. George spends twice as much on utilities as he does on food.

    D. George spends the same amount on food and utilities as he does on mortgage.

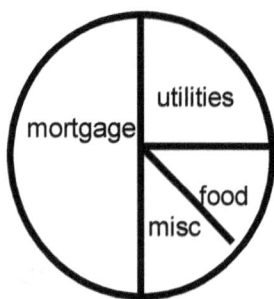

Answer: C: George spends twice as much on utilities as he does on food.

33. Given a spinner with the numbers one through eight, what is the probability that you will spin an even number or a number greater than four?
    (Easy)

    A. 1/4

    B. 1/2

    C. 3/4

    D. 1

Answer: C. 3/4
There are 6 favorable outcomes, 2, 4, 5, 6, 7, 8, and 8 possibilities. Reduce 6/8 to 3/4.

34. **Given a drawer with 5 black socks, 3 blue socks, and 2 red socks, what is the probability that you will draw two black socks in two draws in a dark room?**
    (Average)

    A. 2/9

    B. 1/4

    C. 17/18

    D. 1/18

Answer: A. 2/9
In this example of conditional probability, the probability of drawing a black sock on the first draw is 5/10. It is implied in the problem that there is no replacement, therefore the probability of obtaining a black sock in the second draw is 4/9. Multiply the two probabilities and reduce to lowest terms.

35. **If a horse will probably win three races out of ten, what are the odds that he will win?**
    (Rigorous)

    A. 3:10

    B. 7:10

    C. 3:7

    D. 7:3

Answer: C. 3:7
There are 3 slots to fill. There are 3 choices for the first, 7 for the second, and 6 for the third. Therefore, the total number of choices is 3(7)(6) = 126.

36. Find the median of the following set of data:

    14 3 7 6 11 20

    (*Average*)

    A.   9

    B.   8.5

    C.   7

    D.   11

Answer: A. 9
Place the numbers is ascending order: 3 6 7 11 14 20. Find the average of the middle two numbers: (7+11)/2 = 9.

37. Corporate salaries are listed for several employees. Which would be the best measure of central tendency?
    (*Average*)

    $24,000 $24,000 $26,000
    $28,000 $30,000 $120,000

    A.   Mean

    B.   Median

    C.   Mode

    D.   No difference

Answer: B. Median
The median provides the best measure of central tendency in this case where the mode is the lowest number and the mean would be disproportionately skewed by the outlier $120,000.

38. **A student scored in the 87th percentile on a standardized test. Which would be the best interpretation of his score?**
    (Easy)

    A.	Only 13% of the students who took the test scored higher.

    B.	This student should be getting mostly Bs on his report card.

    C.	This student performed below average on the test.

    D.	This is the equivalent of missing 13 questions on a 100-question exam.

Answer: A. Only 13% of the students who took the test scored higher.
Percentile ranking tells how the student compared to the norm or the other students taking the test. It does not correspond to the percentage answered correctly but can indicate how the student compared to the average student tested.

39. **Determine the number of subsets of set *K*.**

    K = {4, 5, 6, 7}

    (Average)

    A.	15

    B.	16

    C.	17

    D.	18

Answer: B. 16
A set of n objects has n^2 subsets. Therefore, here we have $4^2 = 16$ subsets. These subsets include four which each have 1 element only, six which each have 2 elements, four which each have 3 elements, plus the original set, and the empty set.

40. {6, 11, 16, 21...}

Find the sum of the first 20 terms in the sequence.
(Rigorous)

A. 1070

B. 1176

C. 969

D. 1069

Answer: A. 1070

Apply the formula $\frac{n}{2}[2a_1 + (n-1)d]$ where n = 20, a_1 =6 and d=5.

ANSWER KEY

1.	A	21.	B
2.	B	22.	A
3.	D	23.	D
4.	C	24.	C
5.	A	25.	C
6.	D	26.	B
7.	D	27.	B
8.	C	28.	C
9.	C	29.	C
10.	C	30.	C
11.	A	31.	D
12.	C	32.	C
13.	C	33.	C
14.	B	34.	A
15.	D	35.	C
16.	B	36.	A
17.	C	37.	B
18.	C	38.	A
19.	D	39.	B
20.	A	40.	A

RIGOR TABLE

	Easy 20%	Average 40%	Rigorous 40%
Question	5, 10, 16, 21, 23, 32, 33, 38	1, 2, 4, 7, 8, 11, 14, 17, 25, 26, 29, 30, 34, 36, 37, 39	3, 6, 9, 12, 13, 15, 18, 19, 20, 22, 24, 27, 28, 31, 35, 40